Questions in Higher Human Biology

Text © 2003 Andrew Morton
Design and layout 2003 Leckie & Leckie Ltd
Cover image © SCIENCE PHOTO LIBARY

08/150109

ISBN 978-1-898890-17-1

Published by
Leckie & Leckie Ltd
3rd Floor, 4 Queen Street, Edinburgh EH2 1JE
tel: 0131 220 6831   fax: 0131 225 9987
email: enquiries@leckieandleckie.co.uk   web: www.leckieandleckie.co.uk

Special thanks to
Julie Barclay (design), Lyall Forbes (content review) and Priscilla Sharland (content review)

Leckie & Leckie makes every effort to ensure that all paper used in our books is made from wood pulp obtained from well-managed forests, controlled sources and recycled wood or fibre.

A CIP Catalogue record for this book is available from the British Library.

Leckie & Leckie Ltd is a division of Huveaux plc.

Andrew Morton

Contents

Introduction

This book is designed to be used as a companion study guide to Leckie & Leckie's *Higher Human Biology Course Notes*, written by the same author.

Its aim is to allow you to revise effectively for the Higher Human Biology examinations offered by the SQA every year. At the start of each section, you will find a simple syllabus summary which tells you what you ought to know for the examination. At the foot of each page of questions you will find a page reference to *Higher Human Biology Course Notes*. As you work through the questions you can look up these pages or refer to the extensive index at the end of this book. Topics not covered by questions at the beginning of the book are often covered by questions in the specimen paper.

The pull-out pages at the back of the book provide answers. If this book is provided by your school or college, these pages may have been removed by your teacher or lecturer.

The Higher Human Biology course is divided into three units:

Unit 1 Cell Function and Inheritance
Unit 2 The Continuation of Life
Unit 3 Behaviour, Populations and the Environment

At the end of each unit you must sit a short forty-five minute test of 40 marks. The pass mark for each test is 26 (65%). If you fail a unit test you will be given at least one further opportunity to pass a different unit test of the same standard. You must also carry out, and write up, one practical investigation. Your practical investigation and unit tests will be marked by your teacher or lecturer and can be taken at any time during the year.

At the end of the course, you sit an extended course examination paper which tests your knowledge of all three units in an integrative way. There are also questions in the examination which test your ability to deal with data and to solve problems. The course examination will be more difficult than the unit tests and the pass mark will normally be fixed at around 50%. If you pass, you will be graded 'A', 'B' or 'C' according to your mark. To be sure of an 'A' pass, you should aim to achieve 75% or more in the course examination.

The course examination is $2\frac{1}{2}$ hours long and comprises three sections giving a total of 130 marks.

Section A 30 multiple-choice questions   30 marks
Section B Short-answer questions   80 marks
Section C Two long-answer (essay) questions with a choice   20 marks

A full specimen examination paper with answers is included in this book.

Unit 1 – Cell Function and Inheritance

CELL STRUCTURE AND FUNCTION

- **Know the structure and function of the following organelles:**
  nucleus, nucleolus, mitochondrion, Golgi body, endoplasmic reticulum (E.R.), ribosome, lysosome.

1. Complete the key to enable the cell organelles listed above to be identified.

   1 { organelle with membranes   2
       { organelle without membranes   3

   2 { organelle composed of layers or stacks of membranes   4
       { organelle spherical or lozenge-shaped   5

   3 { organelle often found on the surface of E.R.   (i) _____
       { organelle found inside the nucleus   (ii) _____

   4 { organelle associated with many vesicles   (iii) _____
       { organelle with few or no vesicles   (iv) _____

   5 { organelle bounded by two membranes   6
       { organelle bounded by one membrane   (v) _____

   6 { inner membrane folded   (vi) _____
       { inner membrane not folded   (vii) _____

2. Identify each of the following organelles from a description of its function.

   (a) site of Krebs cycle

   (b) site of protein synthesis

   (c) contains powerful digesting enzymes

   (d) site of ribosome synthesis

   (e) contains DNA

   (f) involved in the packaging of complex molecules prior to their secretion

3. Calculate the actual size, in micrometres (µm), of the cells drawn below.
   The magnifications of the drawings are given in brackets.

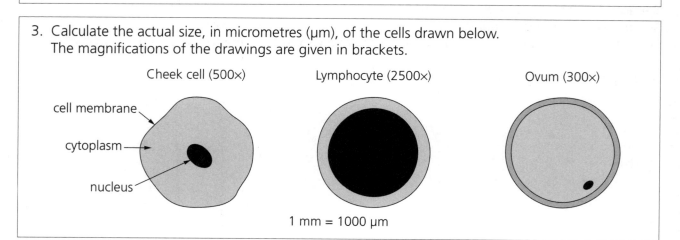

Cheek cell (500×)   Lymphocyte (2500×)   Ovum (300×)

cell membrane

cytoplasm

nucleus

1 mm = 1000 µm

THE ROLE OF ENZYMES IN CELL METABOLISM

- Know that enzymes act as catalysts within and outwith cells.
- Know that the absence of an enzyme blocks a metabolic pathway.
- Be able to describe how enzyme activity is affected by inhibitors, pH, temperature, substrate concentration and enzyme concentration.
- Know that enzymes can be activated by mineral ions, vitamins or by other enzymes.

4. The rate of a reaction catalysed by an amylase enzyme was measured in acidic, alkaline and neutral conditions.
   The results are given in the table below.

pH	Time taken to produce 1 µg of maltose (seconds)
4	55
5	35
6	25
7	20
8	20
9	30
10	50

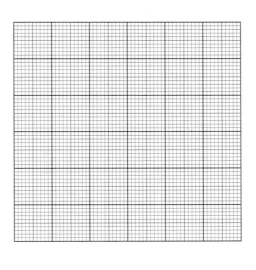

   (a) Plot a line graph to show the relationship of pH to the rate of the reaction.

   (b) From the graph, estimate the optimum pH for the activity of this enzyme.

   (c) What is the substrate of the amylase enzyme?

   (d) At high and low pH the enzyme becomes *denatured*. Explain what is meant by this term.

   (e) Why are vitamins and minerals important constituents of our diet?

   (f) Where are enzymes manufactured in cells?

   (g) Why are many different enzymes present in cells?

5. The metabolic pathway for the production of compound **D** in a cell is shown below.

   $$A \xrightarrow{\text{enzyme } x} B \xrightarrow{\text{enzyme } y} C \xrightarrow{\text{enzyme } z} D \text{ (final product)}$$

   (a) If a mutation occurs, and this mutation alters the shape of the active site of *enzyme y*, predict what will happen to the concentrations of compound **B** and compound **C** present in the cell.

   (b) Excess of compound **D** acts as an inhibitor on *enzyme x*. Why might this be of benefit to the body?

   (c) Name a human disorder caused by the absence of an enzyme in a metabolic pathway.

PROTEIN SYNTHESIS

- Be able to give an elementary description of protein structure.
- Know that enzymes, antibodies and many hormones are proteins.
- Know that proteins are involved in transport and in the structural support of cells and tissues.
- Be able to describe the arrangement and action of actin and myosin fibres in muscle.
- Be able to describe the basic structure of DNA and RNA molecules.
- Be able to describe the role of DNA, mRNA and tRNA in protein synthesis.

---

6. (a) Name the **four** chemical elements always found in protein molecules.

   (b) How many different amino acids are found in nature?

   (c) Name **two** types of bond which bind amino acids in protein molecules.

   (d) What term is used to describe a short chain of amino acids?

---

7. The table shows the locations of four proteins in the body.
   Match the following proteins to their locations.
   actin, haemoglobin, collagen, prolactin

Protein	Location in the body
	red blood cells
	bones, tendons and skin
	muscles
	dissolved in blood plasma

---

8. The following statements relate to the striated appearance of muscle. Which **three** are correct?
   A   The dark bands are composed of myosin only.
   B   The light bands are composed of actin only.
   C   The dark bands are composed of actin and myosin filaments which overlap.
   D   The light bands become narrower when muscles relax.
   E   The dark bands remain the same width when muscles contract.

---

9. The diagram below shows protein synthesis taking place at a ribosome.

   (a) Give the abbreviated names for the next **three** amino acids which would be linked at the ribosome.

   (b) What are the **two** DNA triplet codes for the amino acid *tyrosine* (TYR)?

   (c) What is the maximum number of different triplet codes that it is possible to generate using the four bases A, C, G and T?

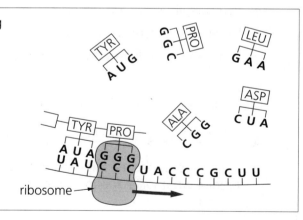

---

10. Complete the table, using ticks, crosses and the names of sugars, to show the differences between DNA and RNA.

Feature	DNA	RNA
double helix		
uracil present		
name of sugar		

ENERGY TRANSFER

- Know that the energy of ATP drives cell metabolism.
- Be able to describe the three stages of aerobic respiration in some detail.
- Know the following chemicals: glucose, pyruvic acid, acetyl CoA, citric acid and NAD.
- Know the number of carbon atoms of compounds at each stage of respiration.
- Be able to distinguish between aerobic and anaerobic respiration.
- Know that carbohydrates can be classified as monosaccharides, disaccharides and polysaccharides.
- Know that carbohydrates, lipids and proteins are sources of energy for the cell.
- Be able to give a brief description of the many roles of lipids.

---

11. Use the letters **G**, **K** and **C** after each statement to indicate whether the statement refers to **G**lycolysis, the **K**rebs cycle or the **C**ytochrome system. In some answers, **two** letters are required.

    (a) takes place in the cytoplasm

    (b) results in the production of NAD(H)

    (c) pyruvic acid is the end product

    (d) results in the direct production of ATP

    (e) takes place in the mitochondria

    (f) ends with the manufacture of water

    (g) occurs during anaerobic respiration

    (h) coenzymes are reduced and oxidised

---

12. How many carbon atoms do each of the following molecules contain?

    (a) glucose   (b) lactic acid   (c) citric acid

---

13. The table below shows a variety of chemical tests which can be used to identify some carbohydrates.

Example	Solubility	Barfoed's reagent	Benedict's reagent	Iodine solution	Clinistix
glucose	+	+	+	–	+
starch	–	–	–	+	–
sucrose	+	–	–	–	–
maltose	+	–	+	–	–
fructose	+	+	+	–	–

+ indicates a positive result   – indicates a negative result

    (a) Which test can be used to distinguish between glucose and fructose?

    (b) From the table, select an example of a monosaccharide, a disaccharide and a polysaccharide.

    (c) Name the soluble carbohydrate which reacts with Benedict's but not with Barfoed's reagent.

    (d) Explain why insoluble compounds such as starch and glycogen are useful as stores of energy.

---

14. Name the parts of the molecule labelled **X** and **Y** in the diagram of a fat (triglyceride) molecule shown opposite.

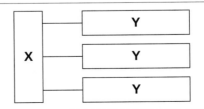

CELL TRANSPORT

- Be able to describe the fluid-mosaic model of cell membranes.
- Know that liquid and gas molecules tend to diffuse from areas of high concentration to areas of low concentration.
- Know that osmosis is the diffusion of water through a membrane.
- Know that active transport is the movement of molecules across a membrane against the concentration gradient using ATP as a source of energy.
- Be able to describe endocytosis and exocytosis.
- Be able to distinguish between pinocytosis and phagocytosis as examples of endocytosis.

15. The diagram is of a cross-section through the plasma membrane of a cell. Glucose molecules Ⓖ are shown crossing from one side of the membrane to the other **against** a concentration gradient.

    (a) What term is used to describe this type of membrane transport?

    (b) What type of molecule is **P**?

    (c) What is the energy source for this type of transport?

    (d) Name **two** other processes by which molecules can cross from one side of a membrane to another.

    (e) What feature of the phospholipids results in their parallel orientation in the membrane?

    (f) Membranes are often said to be *selectively permeable*. What is meant by this term?

    phospholipid

16. Complete the key by naming the methods of movement of substances across a membrane.

    1 { substance moves by virtue of its own kinetic energy   2
        { substance is moved by expenditure of ATP energy   3

    2 { substance is a soluble mineral ion or gas   (i) _____
        { substance is water   (ii) _____

    3 { substance is engulfed by the plasma membrane (endocytosis)   4
        { substance is pumped through membrane by carrier proteins (iii) _____

    4 { substance is solid   (iv) _____
        { substance is liquid   (v) _____

CELLULAR RESPONSE IN DEFENCE

- Know that body cells have unique antigenic molecules in their membranes and that the defence system recognises these as being 'self' antigens.
- Know about the ABO blood groups in relation to 'self' recognition.
- Know that B-lymphocytes produce humoral antibodies in response to foreign antigens.
- Be able to describe the cell-mediated response of T-lymphocytes.
- Be able to describe the function of macrophages in defence.
- Be able to distinguish between artificial, natural, active and passive immunity.
- Know that allergy and autoimmunity are examples of a failure in the normal defence system.
- Be able to describe the nature of viruses and their invasion of cells.

17. The diagram shows the sequence of events when the skin is punctured and bacteria invade the tissues.

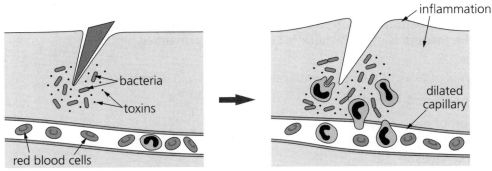

    (a) Name the cells which are engulfing the bacteria.

    (b) What term is used to describe this process?

18. The diagram opposite represents an antibody.

    (a) Which cells of the immune system produce humoral antibodies?

    (b) To what does the binding site attach?

    (c) The binding site is *highly specific*. What does this mean?

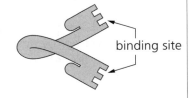

19. Use the terms *active* and *passive,* and *natural* and *artificial* to identify the forms of immunity described below.

    (a) Antibodies are passed across the placenta from mother to child.

    (b) Antibodies are produced after infection.

    (c) Antibodies are produced after vaccination.

20. Distinguish between autoimmunity and allergy.

21. The diagram opposite is of a virus.

    (a) Name the **two** substances labelled **X** and **Y**.

    (b) How many nanometres (nm) are there in a micrometre (μm)?

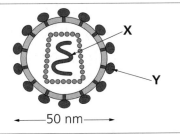

INHERITANCE

- Know that chromosomes carry genes and that genes are composed of DNA.
- Be able to describe DNA replication.
- Know the difference between autosomes and sex chromosomes.
- Be able to describe mitosis and meiosis and distinguish between these two processes.
- Be able to solve simple monohybrid crosses to the F_2 generation.
- Be able to solve genetics problems involving multiple alleles, co-dominant and dominant alleles.
- Know the effect of sex-linkage on inheritance patterns.
- Know that polygenic inheritance leads to a normal distribution of phenotypes.
- Be able to describe different types of mutation.
- Be able to describe non-disjunction and its effects on human karyotypes.
- Know about genetic screening and counselling and the use of family histories in determining genotypes. Examples include albinism, Huntington's chorea, cystic fibrosis, phenylketonuria, haemophilia and muscular dystrophy.
- Be able to describe risk evaluation in relation to polygenic inheritance.
- Know that post-natal screening is used to check for genetic abnormalities.

---

22. The diagram below is of a section of deoxyribonucleic acid (DNA) during replication.

    (a) Identify the bases **1** to **6**.

    (b) What proportion of each new replicated strand comes from the original strand of DNA?

    (c) Name **two** compounds required by the cell for replication to take place.

    (d) What term is used to describe a piece of DNA which codes for a protein?

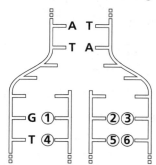

---

23. (a) The list below describes various stages in meiosis.
    Using the letters, place the stages in the order in which they would occur.
    A   chromosomes find their homologous partners
    B   chromosomes shorten and thicken
    C   pairs of chromosomes migrate to the equator of the cell
    D   the cells divide to form four haploid cells
    E   the original cell divides in two
    F   chromosomes migrate to the poles of the cell
    G   single chromosomes migrate to the equators of the cells
    H   chromatids migrate to the poles of the cells
    I   chiasmata form

    (b) Choose any **two** statements from the list above which confirm that the process described is *meiosis* and not *mitosis*.

---

24. The diagram shows a pair of homologous chromosomes during meiosis.

    (a) Name parts **X**, **Y** and **Z**.

    (b) Why is the exchange of genetic material at **X** important?

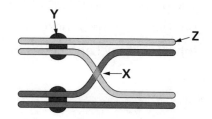

---

25. Polydactyly is a dominant genetic disorder which results in the development of hands with more than five digits. The diagram shows a family tree in which the condition has been inherited from the grandfather.

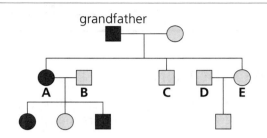

grandfather

□ normal male ■ polydactyl male
○ normal female ● polydactyl female

(a) How many children and grandchildren do the grandparents have?

(b) (i) Is polydactyly a sex-linked condition?
(ii) What information in the family tree allows you to reach this conclusion?

(c) Using the letters **P** for the dominant allele and **p** for the recessive normal allele, give the genotypes of individuals **A** and **B**.

26. Albinism is a genetically recessive characteristic, the allele for which is present in 1 in 70 of the population. Answer the following questions using the letters **M** to represent the allele for normal pigmentation and **m** to represent the recessive allele.

(a) What is the genotype of an albino?

(b) Which pairing of individuals listed below could be the parents of an albino child?
Mm × Mm, MM × mm, MM × Mm

(c) An albino woman marries a phenotypically normal male of unknown genotype and they have a child. What are the chances of the child being an albino?

27. The **MN** blood group system shows co-dominance. Heterozygotes produce both **M** and **N** blood group antigens. The **MN** blood groups do not have to be matched for transfusions.

(a) A man and woman, both with blood group **MN**, have a child with blood group **M**. Draw a Punnett diagram to show how this arises.

(b) The couple expect another child. What are the chances of it too being of blood group **M**?

28. Complete the following sentence by underlining the correct alternatives.
In polygenic inheritance, characteristics are controlled by **one gene/many genes** and give rise to **discontinuous/continuous** variation.

29. Complete the following description of mutation, choosing appropriate words from the list below. (*Not all the words are used, but words may only be used once.*)

random, cell, chromosome, autosome, gamete, base, Down's syndrome, mutagen, translocation, insertion, non-disjunction, chiasmata, DNA, cystic fibrosis

Mutations can occur in any _____. However, only a mutation occurring in a _____ is of any real significance, because it is inherited. Mutations are _____, undirected, spontaneous changes in _____ molecules and can be as simple as a change in one _____, or as significant as a change to an entire _____. During meiosis a mutation called _____ can result in extra chromosomes in a gamete. This is often lethal but with _____ number 21 it results in a condition called _____ _____.

Unit 2 – The Continuation of Life

REPRODUCTION

- Be able to describe the basic structure and function of the reproductive organs, including the seminiferous tubules and interstitial cells of the testes, and the Graafian follicle and corpus luteum of the ovaries.
- Know about the contribution to fertilisation of the secretions of the prostate gland and the seminal vesicles.
- Be able to describe the hormonal control of the menstrual cycle.
- Know about the influence of LH and FSH on the activity of the testes.
- Be able to describe different causes and treatments of infertility in males and females.
- Know about forms of contraception which are biologically based, e.g. the rhythm method and the contraceptive pill.

---

30. The diagrams show side views of the female and male reproductive organs.

Female 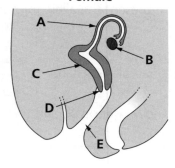   **Male**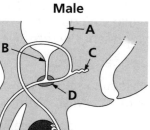

Which letters label the following structures?

Female: vagina
       cervix
       uterus
       oviduct
       ovary

Male: prostate gland
      seminal vesicle
      testis
      bladder
      urethra

---

31. The graph shows changes in concentrations of the hormones LH, oestrogen and progesterone in the blood of a woman during a complete oestrous cycle.

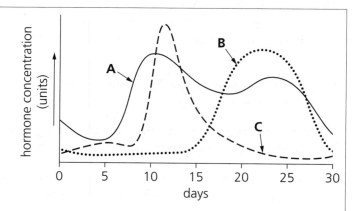

(a) Identify the hormones.

(b) For each hormone, state where it is produced.

(c) Which **two** of these hormones are often used in contraceptive pills?

(d) During which days of the cycle is the woman most likely to be fertile?

DEVELOPMENT

- Be able to describe cleavage, implantation and differentiation of the zygote.
- Be able to describe, in brief, pre-natal and post-natal development.
- Be able to explain how monozygotic and dizygotic twins arise.
- Know about the exchanges between maternal and fetal circulations across the placenta.
- Know about the influences of the hormones progesterone, oestrogen and prolactin.
- Be able to explain the effects of the fetus on the maternal immune system with reference to the Rhesus antigen.
- Know about the role and artificial use of oxytocin at birth.
- Know about the nutrition of the newborn.
- Be able to describe the major stages of growth after birth and the role of growth hormone in the control of growth and development.
- Know the major body changes and hormonal changes which occur at puberty.

---

32. Give the terms for each of the following definitions.

    (a) a fertilised egg

    (b) twins which arise from a single fertilised egg

    (c) the first few divisions of a fertilised egg

    (d) the lining of the uterus

---

33. (a) Use the names of the hormones in the list below to answer the questions which follow. (*Each name can be used more than once.*)

    progesterone, oestrogen, prolactin, oxytocin

      (i) Which **two** hormones are produced by the placenta?
      (ii) Which **two** hormones are produced by the pituitary gland?
      (iii) Which hormone stimulates the production of milk by the breasts?
      (iv) Which hormone stimulates the ejection of milk from the breasts?
      (v) Which hormone stimulates the contraction of the uterus?

    (b) What effect does suckling at the breast have on the hypothalamus?

    (c) (i) What name is given to the milk first produced by the breast shortly after birth?
       (ii) Describe **two** ways in which this milk is different from breast milk produced later.

---

34. The graph shows the percentage of total growth of different parts of the body between birth and age 19.

    (a) Why is it important for the brain to complete most of its growth at an early stage?

    (b) (i) What term is used to describe the stage at which the sexual organs start to become mature?
       (ii) From the graph, at what age does this stage begin?
       (iii) For each sex, name a hormone which promotes maturation of the sex organs.

    (c) Name the hormone which promotes the growth of bones.

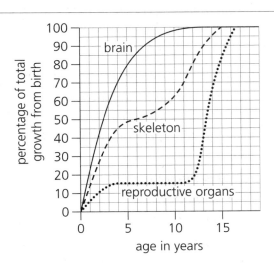

TRANSPORT MECHANISMS

- Be able to distinguish between plasma, tissue fluid and lymph.
- Be able to describe the differences between arteries, veins and capillaries.
- Know the relationship of arterioles, venules, capillaries, and lymphatic capillaries to the body cells and tissue fluid.
- Know about the exchange of materials between the body cells and tissue fluid.
- Be able to identify the chambers and valves of the heart, and the main blood vessels of the body.
- Be able to describe the cardiac cycle and the control of heart rate.
- Be able to explain changes in blood pressure as blood flows round the body.
- Know the basic features of the lymphatic circulation and lymph nodes.

---

35. Use the names of the blood vessels given below to answer the questions which follow. *(Each answer may be used once, more than once, or not at all.)*

    artery, arteriole, capillary, vein

    (a) Which one pulsates?

    (b) Which one contains valves?

    (c) In which one does blood pressure drop most quickly?

    (d) Which one has thick muscular walls?

    (e) Which one has permeable walls?

    (f) Which one carries oxygenated blood in the umbilical cord?

---

36. The graph shows the pressure changes which take place in the **left** side of the heart during one complete cardiac cycle.

    (a) Name the valves which close at points **X** and **Y**.

    (b) Between what **two** times does atrial systole take place?

    (c) At what time does ventricular systole start?

    (d) Between what **two** times is the pressure in the ventricle higher than in the aorta?

    (e) Explain why blood pressure in the right side of the heart is lower than in the left side.

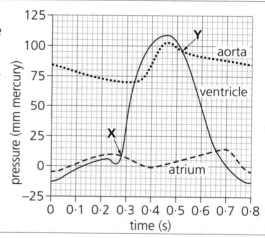

---

37. The diagram shows a lymph gland (node).

    (a) What cells of the immune system are particularly common in such glands?

    (b) Why do lymph glands often swell up during illness?

    (c) What is the source of the lymph which enters these glands?

    (d) To where does the lymph go after it leaves the gland?

    (e) In which body fluid is most protein found?
    plasma, tissue fluid, lymph

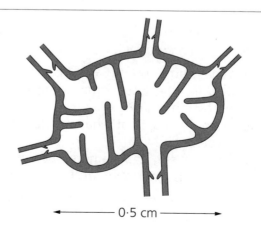

0·5 cm

THE DELIVERY AND REMOVAL OF MATERIALS

- Know about the affinity (attraction) of haemoglobin for oxygen under different conditions.
- Know that the lungs excrete carbon dioxide.
- Be able to describe the structure and function of red blood cells.
- Know the life history of a red blood cell.
- Be able to describe how glucose, amino acids and vitamin B_{12} are absorbed through villi in the gut.
- Know about the role of bile salts and lacteals in the digestion and absorption of lipids.
- Know that the liver has a unique dual blood supply.
- Be able to describe, in brief, the many functions of the liver.
- Know the fate of absorbed materials such as carbohydrates, proteins, lipids, vitamins and minerals (covered mostly in Unit 1).
- Be able to describe the function of the kidney nephron.

38. Take your answers for the following questions from the graph on page 42 of *Higher Human Biology Course Notes.*

    (a) Complete the following table by taking partial pressure (pp) readings from the graph. The first line of the table is completed for you.

	% oxygen in blood at pp of 40 mmHg	% oxygen in blood at pp of 30 mmHg	% of oxygen released
blood without haemoglobin	38	28	38 – 28 = 10
haemoglobin at normal temperature			

    (b) From the table above, what effect does the presence of haemoglobin have on the ability of blood to release oxygen at respiring tissues?

    (c) At a partial pressure of 30 mmHg, how much more oxygen is released when the temperature rises from normal to the higher temperature shown on the graph?

39. (a) State **two** features of red blood cells which increase their surface area to volume ratio.

    (b) Name the breakdown product of haemoglobin which gives bile its characteristic colour.

40. Name the blood vessels which run between the following organs and blood vessels and state whether each vessel carries oxygenated or deoxygenated blood.

    (a) from intestines to liver

    (b) from aorta to liver

    (c) from aorta to kidney

    (d) from heart to lungs

41. Complete the description of deamination using appropriate words.

    Deamination takes place in the _____. During this process excess _____ _____ are converted to keto acids and _____. The _____ is very toxic and is immediately converted to _____ which is then carried to the _____ where it is excreted. The keto acids can be stored as _____ or oxidised to produce _____.

REGULATING MECHANISMS

- Be able to explain the principles of negative feedback mechanisms.
- Know about the effect of exercise on the cardiovascular and respiratory systems.
- Know the roles of the hormones insulin, glucagon and adrenaline in the control of blood sugar concentrations.
- Be able to describe how body temperature is maintained by voluntary and involuntary responses.
- Be able to calculate surface area to volume ratios and relate this to heat gain and heat loss.
- Know that the hypothalamus monitors blood temperature.
- Be able to explain why the very young and the very old are more susceptible to hypothermia.

---

42. The diagram to the right shows stages in the control of body temperature.

    (a) Where are changes in body temperature detected in the body?

    (b) Name one corrective mechanism which will be switched on.

    (c) Describe how this corrective mechanism brings body temperature down.

    (d) What term is used to describe this homeostatic mechanism?

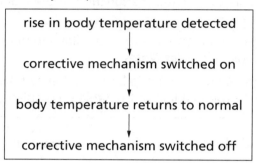

rise in body temperature detected

↓

corrective mechanism switched on

↓

body temperature returns to normal

↓

corrective mechanism switched off

---

43. The graphs show the volumes of air breathed in and out by a runner before starting, and immediately after, a race.

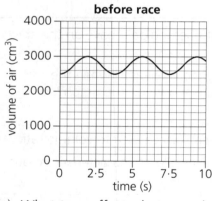

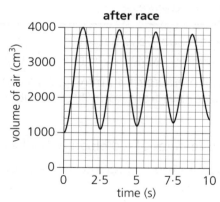

    (a) What **two** effects does exercise have on breathing?

    (b) Before the race, what volume of air is breathed every **minute**?

    (c) Which gas stimulates the breathing centres of the brain?

---

44. The table below shows some of the hormones which influence sugar concentrations in the blood. Which line of the table is correct?

Line	Hormone	Site of production	Action
1	adrenaline	pituitary gland	raises blood sugar
2	insulin	liver	lowers blood sugar
3	glucagon	pancreas	raises blood sugar

See pages 42, 45 and 46 of Leckie & Leckie's *Higher Human Biology Course Notes*.   © Leckie & Leckie

Unit 3 – Behaviour, Populations and the Environment

THE NERVOUS SYSTEM

- Be able to describe the basic structure and function of the brain, including the cerebrum, cerebellum, medulla oblongata and corpus callosum.
- Be aware that humans have a relatively large brain which gives much greater computing power than that of any other animal on earth.
- Know that many functions are localised in the cerebrum, which is convoluted (folded) to increase its surface area.
- Know that there is a relationship between the size of a discrete region in the brain and the function it carries out.
- Know that the nervous system is divided into central, peripheral, somatic and autonomic parts.
- Know the roles of the sympathetic and parasympathetic sub-divisions of the autonomic nervous system.
- Know the basic structure of motor and sensory neurones, including the cell body, dendrites, axon and myelin sheath.
- Know the importance of myelination of nerve fibres and that myelination continues after birth.
- Be able to describe the chemical transmission of nervous impulses at synapses.
- Know about the action of acetylcholine and noradrenaline as neurotransmitters.
- Be able to distinguish between converging and diverging neural pathways.

---

45. The diagram shows a side view of part of the brain.

    (a) What part of the brain is shown?

    (b) Why is the surface of the brain folded many times?

    (c) What functions are controlled by the motor area?

    (d) Name the strip which lies next to the motor area.

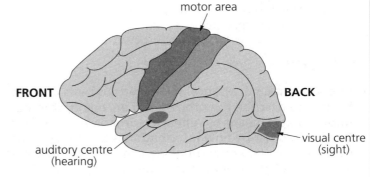

    (e) The part of the brain shown is divided into two halves. How are these halves linked?

---

46. The diagram below shows the major divisions of the nervous system.

    (a) Complete the blanks.

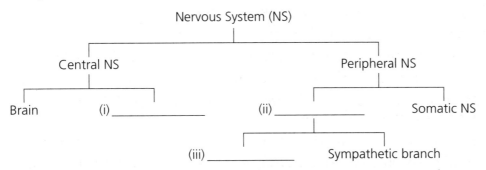

    (b) Under what circumstances does the sympathetic system come into play?

    (c) What effect does the parasympathetic system have on blood distribution to the digestive system?

---

47. The diagram shows a neurone.
    (a) What type of neurone is shown?
    (b) Name the **four** parts labelled.
    (c) In which direction (left to right or right to left) would an impulse travel along this neurone?

48. The following questions relate to synaptic transmission.
    (a) How are neurotransmitter chemicals transported to the membrane of the axon?
    (b) How does the neurotransmitter cross the synapse?
    (c) What effects can neurotransmitters have on their receptors?
    (d) What happens if insufficient neurotransmitter is produced?
    (e) What happens to neurotransmitters after they have reacted with their receptor?

49. The diagram shows cells of the retina and their nerve connections.
    (a) Where is the retina located in the body?
    (b) What types of cell are shown?
    (c) Name the type of nerve pathway shown in the diagram.
    (d) Explain how this arrangement of nerves increases sensitivity.

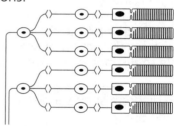

MEMORY

- Know that memory is localised in various parts of the brain. One such part is the limbic system.
- Be able to describe the processes of encoding, storage and retrieval of memories.
- Know what is meant by rehearsal, organisation and elaboration of meaning, and contextual cues.
- Be able to describe how information is transferred from short-term to long-term memory.
- Be able to explain how chunking can increase the capacity of short-term memory.
- Be able to explain what causes the serial position effect.
- Know that the absence of neurotransmitters, such as acetylcholine, affects memory.
- Know that the presence of the receptor NMDA in the limbic system suggests that NMDA is important in memory storage.

50. (a) For memory to function, memories have to be placed in the brain, kept there and be recoverable. What **three** terms describe these processes?
    (b) Name **one** part of the brain where memories are stored.
    (c) Name **two** methods of encoding information to form a memory.
    (d) How many pieces of information can be stored in short-term memory (STM)?
    (e) What term is used to describe the memory capacity of STM?
    (f) (i) What technique can be used to increase the storage capacity of STM?
        (ii) Give an example of this technique.
    (g) What happens to information in STM once its full capacity has been reached?
    (h) In what part of the brain is the learned ability to ride a bike or catch a ball stored?
    (i) (i) What is a mnemonic? (ii) Give an example of a mnemonic.
    (j) When retrieving information from the brain, contextual cues are sometimes used. Give **two** examples of contextual cues.

BEHAVIOUR

- Know that genes, environment and stage of maturation all influence behaviour.
- Be aware that the sequential stages of the development of skills such as walking are fairly constant and fixed.
- Know that phenylketonuria (PKU) and Huntington's chorea are examples of inherited disorders which affect behaviour.
- Be able to explain the value of studies of twins in investigating the influence of the environment on behaviour.
- Know that intelligence is an example of a brain function which is influenced by many genes, the environment and the age of a person.
- Be aware that humans have a very long period of child dependency to allow time to learn.
- Know about the importance of infant attachment in the development of later competent and acceptable social behaviour.
- Be aware of the significance of verbal and non-verbal communication as part of human behaviour.
- Know about the importance of imitation and practice in the development of many behaviours and skills.
- Be able to distinguish between reinforcement, shaping and extinction as seen in trial and error learning.
- Be able to distinguish between generalisation and discrimination behaviours.
- Know the effects of group behaviour and social influence on the behaviour of the individual.
- Be able to distinguish between social facilitation, deindividuation, internalisation and identification as types of behaviour influenced by others.

---

51. Three facial expressions are shown opposite.

    (a) State what meaning each expression conveys.

    (b) Describe how the arms or hands might be used to emphasise expression **C**?

    (c) What term is used to describe communication which does not involve speech?

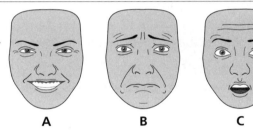

A　　　**B**　　　**C**

---

52. The graph shows the average age at which children start to use words and compose sentences.

    (a) What percentage of one-year-old children can say recognisable words?

    (b) At what age would 50% of children be expected to be able to compose sentences?

---

53. The illustration shows various stages reached by young children as they learn to walk.

    (a) What **three** factors influence the development of the ability to walk?

    (b) What happens to nerve axons in the body as these stages progress?

    (c) Suggest the next stage in learning to walk.

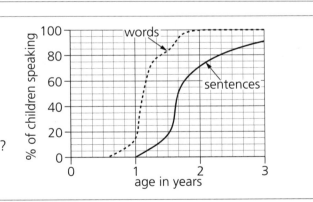

---

54. (a) Why are identical twins of particular value in the study of human behaviour?

    (b) Why is the period of human dependency so long?

55. Select terms from the list below which could be used to describe the following situations. *(Each answer may be used only once.)*

    discrimination, identification, shaping, imitation, generalisation, reinforcement, practice, instinct

    (a) A boy learns how to hold a snooker cue by watching an expert play.

    (b) A girl learns to play the piano by playing melodies over and over each day.

    (c) A boy becomes scared of all flying insects because he has been stung by a wasp.

    (d) A girl learns that nettles sting and avoids them thereafter.

    (e) A boy wants to play tennis to be like his father, who plays tennis well.

POPULATION GROWTH

- Understand the concept of carrying capacity as it relates to early and existing human populations.
- Know that the rise of any population can be exponential if left unchecked.
- Know that disease and lack of food and water are important factors in limiting population growth.
- Be able to distinguish demographic trends in developed and developing countries.

56. Populations have the potential to grow exponentially if left unchecked.

    (a) State **two** factors which can check the growth of populations.

    (b) What is meant by the term *exponential*?

    (c) What development took place some ten thousand years ago, which had a dramatic effect on the carrying capacity of the Earth?

57. **Graph 1** shows the projected population changes in the UK for four different fertility rates.

    Fertility rate is the average number of children per woman in the population. It can be calculated by multiplying the number of families of a particular size by the number of children in that size of family, and then dividing the total by the overall number of families.

    For example, in a sample of 20 families, if there are 8 families with 2 children and 12 families with 1 child, the fertility rate is:

    $$\frac{(8 \times 2) + (12 \times 1)}{20} = \frac{28}{20} = 1 \cdot 4$$

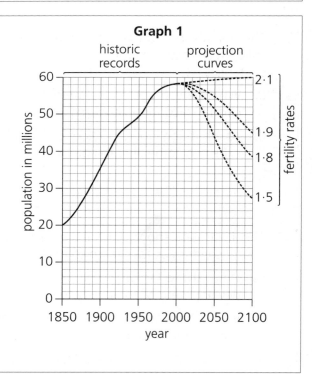

Graph 1

Answers to Units 1, 2 and 3

Solidus (/) indicates an alternative answer.

1. (i) ribosome (ii) nucleolus (iii) Golgi body
    (iv) endoplasmic reticulum (v) lysosome
    (vi) mitochondrion (vii) nucleus

2. (a) mitochondrion
    (b) ribosome
    (c) lysosome
    (d) nucleolus
    (e) nucleus *(The mitochondria also contain DNA but this is beyond the syllabus.)*
    (f) Golgi body

3. Cheek cell: 60 µm
    Lymphocyte: 12 µm
    Ovum: 100 µm

4. (a)

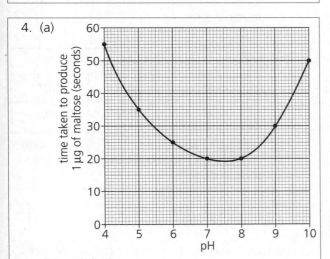

    (b) 7·5 (±2)
    (c) starch *(knowledge required from Standard Grade or Intermediate 2 Biology)*
    (d) The shape of the active site becomes altered.
    (e) They act as activators to allow enzymes to function properly.
    (f) at ribosomes
    (g) because there are many hundreds of different reactions occurring in cells, and each enzyme is highly specific, i.e. it can only catalyse one reaction

5. (a) The concentration of **B** will increase and the concentration of **C** will decrease.
    (b) The body will reduce the production rate of **D** when sufficient has been produced, hence saving raw materials and energy.
    (c) phenylketonuria *(There are many others, but this disorder is specifically mentioned in the syllabus.)*

6. (a) carbon, oxygen, hydrogen and nitrogen
    (b) 20
    (c) hydrogen bonds and peptide bonds
    (d) peptide

7.

Protein	Location in the body
haemoglobin	red blood cells
collagen	bones, tendons and skin
actin	muscles
prolactin	dissolved in blood plasma

8. B, C and E

9. (a) TYR – PRO – LEU
    (b) ATA and ATG
    (c) 64

10.

Feature	DNA	RNA
double helix	✓	✗
uracil present	✗	✓
name of sugar	deoxyribose	ribose

11. (a) G (b) K *(strictly, also G)*
    (c) G (d) G and C
    (e) K and C (f) C
    (g) G (h) C

12. (a) 6 (b) 3 (c) 6

13. (a) Clinistix
    (b) monosaccharide: glucose or fructose
        disaccharide: sucrose or maltose
        polysaccharide: starch
    (c) maltose
    (d) They do not have an osmotic effect on cells./They cannot diffuse out of the cells in which they are stored.

14. **X**: glycerol **Y**: fatty acid

15. (a) active transport
    (b) protein
    (c) ATP
    (d) *Any two from:* endocytosis/exocytosis/
        diffusion/osmosis/phagocytosis/pinocytosis
    (e) The phospholipid molecules have
        hydrophobic and hydrophilic ends. The
        hydrophilic end is attracted to water on the
        outside, while the hydrophobic end is
        repelled and turns inwards.
    (f) The membrane allows some substances to
        pass through and prevents other
        substances from passing through.

16. (i) diffusion
    (ii) osmosis
    (iii) active transport
    (iv) phagocytosis
    (v) pinocytosis

17. (a) macrophages
    (b) phagocytosis

18. (a) B-lymphocytes
    (b) an antigen
    (c) The shape of the binding site is such that it
        can bind with only one type of antigen.

19. (a) passive, natural
    (b) active, natural
    (c) active, artificial

20. Both are failures of the immune system.
    Autoimmunity results in the destruction of 'self'
    body cells and allergy results in over-reaction to
    relatively harmless foreign antigens.

21. (a) **X**: nucleic acid (DNA or RNA)
        Y: protein (*strictly* **glyco***protein*)
    (b) 1000

22. (a) 1: cytosine   2: guanine
         3: cytosine   4: adenine
         5: thymine   6: adenine
    (b) half
    (c) *Any two from:* ATP/enzymes/nucleotides
    (d) a gene

23. (a) B, A, I, C, F, E, G, H, D
    (b) *Any two from:* A, C, D or I

24. (a) **X**: chiasma
        Y: centromere
        Z: chromatid
    (b) It results in variation in gametes and hence
        in individuals.

25. (a) 3 and 4
    (b) (i) no
        (ii) If it were sex-linked, female **E** would
             have polydactyly because she would
             have inherited an X-chromosome, with
             the dominant polydactyly allele on it,
             from her father.
    (c) **A: Pp   B: pp**

26. (a) mm
    (b) Mm × Mm
    (c) 1 in 140 *(There is a 1 in 70 chance that the
        man carries the allele and, if he does, a 1 in
        2 chance of an albino child. The odds are
        multiplied.)*

27. (a)

	M	*N*
M	**MM**	**MN**
N	**MN**	**NN**

    (b) 1 in 4 *(The fact that the 1st child was **MM**
        has no effect on the genotype of the 2nd
        child.)*

28. many genes and continuous

29. Mutations can occur in any *cell*. However, only
    a mutation occurring in a *gamete* is of any real
    significance, because it is inherited. Mutations
    are *random*, undirected, spontaneous changes in
    DNA molecules and can be as simple as a
    change in one *base*, or as significant as a change
    to an entire *chromosome*. During meiosis a
    mutation called *non-disjunction* can result in extra
    chromosomes in a gamete. This is often lethal
    but with *autosome* number 21 it results in a
    condition called *Down's syndrome*.

30. vagina: E   prostate gland: D
    cervix: D   seminal vesicle: C
    uterus: C   testis: E
    oviduct: A   bladder: A
    ovary: B   urethra: B

31. (a) **A**: oestrogen
B: progesterone
C: LH
(b) oestrogen by follicle, progesterone by corpus luteum, LH by pituitary gland
(c) oestrogen and progesterone
(d) between 12th and 15th day approximately

32. (a) zygote   (b) monozygotic
(c) cleavage   (d) endometrium

33. (a) (i) progesterone and oestrogen
(ii) prolactin and oxytocin
(iii) prolactin
(iv) oxytocin
(v) oxytocin
(b) It causes the hypothalamus to stimulate the pituitary to produce prolactin and oxytocin.
(c) (i) colostrum
(ii) *Any two from:* It has more antibodies, protein, minerals and vitamin A. It has less lactose and less fat.

34. (a) The brain has to become fully functional at an early age as much learning has to take place.
(b) (i) puberty
(ii) 11 years
(iii) males: testosterone
females: oestrogen
(c) growth hormone

35. (a) artery   (b) vein
(c) arteriole   (d) artery
(e) capillary   (f) vein

36. (a) **X**: bicuspid   **Y**: semilunar
(b) 0 and 0·24 seconds
(c) 0·26 seconds
(d) between 0·28 and 0·52 seconds
(e) The right hand side of the heart is less muscular because it only pumps blood round the lungs and not the entire body.

37. (a) macrophages and lymphocytes
(b) because of the increased activity of cell production
(c) blind-ended lymph vessels which collect tissue fluid
(d) via a lymphatic vessel to a vein near the heart
(e) plasma

38. (a)

	% oxygen in blood at pp of 40 mmHg	% oxygen in blood at pp of 30 mmHg	% of oxygen released
blood without haemoglobin	38	28	10
haemoglobin at normal temperature	73 ±1	56 ±1	17 ±2

(b) It allows more oxygen to be released.
(c) 13% (±2%)

39. (a) their small size and their biconcave shape
(b) bilirubin

40. (a) hepatic portal vein – deoxygenated
(b) hepatic artery – oxygenated
(c) renal artery – oxygenated
(d) pulmonary artery – deoxygenated

41. Deamination takes place in the *liver*. During this process excess *amino acids* are converted to keto acids and *ammonia*. The *ammonia* is very toxic and is immediately converted to *urea* which is then carried to the *kidneys* where it is excreted. The keto acids can be stored as *glycogen* or oxidised to produce *ATP*.

42. (a) hypothalamus
(b) vasodilation or sweating
(c) Vasodilation brings warm blood nearer to the surface of the skin where heat is lost more easily. Sweating results in the evaporation of water which takes heat energy from the skin.
(d) negative feedback

43. (a) Breathing is quicker and deeper.
(b) 8 litres (1 litre every 7·5 seconds)
(c) carbon dioxide

44. line 3

45. (a) cerebrum/(left) cerebral hemisphere
(b) to increase the surface area for nerves and nerve connections
(c) all conscious muscle movements
(d) sensory area
(e) The two cerebral hemispheres are connected by a band of fibres called the corpus callosum.

46. (a) (i) spinal cord   (ii) autonomic NS
       (iii) parasympathetic branch
    (b) at times of stress, excitement or exercise
    (c) It increases the blood supply to the digestive system.

47. (a) motor neurone
    (b) **A**: dendrite   **B**: nerve cell body
        C: axon   **D**: myelin sheath
    (c) from left to right

48. (a) They are carried in vesicles.
    (b) by diffusion
    (c) They can inhibit or stimulate receptors.
    (d) No stimulus is transmitted.
    (e) They are reabsorbed or removed by the action of enzymes.

49. (a) in the eye   (b) rods (and neurones)
    (c) converging
    (d) Weak stimuli, which would have no effect on their own, are added to make a sufficiently strong stimulus to reach the brain. In this way, very weak light can be detected by rods.

50. (a) encoding → storage → retrieval
    (b) limbic system (*In truth, many parts of the brain are involved in memory storage.*)
    (c) *Any two from:* acoustic (sounds), visual (sights), semantic (meanings) coding
    (d) approximately seven
    (e) memory span
    (f) (i) chunking
        (ii) telephone numbers or car registration numbers. For example, 01312206831 can be chunked into 01-31-22-0-6831.
    (g) It is displaced.
    (h) cerebellum
    (i) (i) a memory aid
        (ii) e.g. remembering **v**eins have **v**alves and **t**RNA has an**t**i-codons
    (j) a **sight**, **smell** or **sound** (*any two*) which aids in bringing back a distant memory

51. (a) **A**: happiness   **B**: sadness   **C**: surprise
    (b) surprise: hands in air/hands on hips
    (c) non-verbal communication

52. (a) 15%
    (b) 1 year 8 months

53. (a) age/maturity, inheritance/genes, environment/surroundings
    (b) They become increasingly myelinated.
    (c) standing unaided or walking with support

54. (a) They are the same age and have the same genotype. So any differences between them must result from environmental factors only.
    (b) Humans have to learn much more than any other animal.

55. (a) imitation   (b) practice
    (c) generalisation   (d) discrimination
    (e) identification

56. (a) disease and lack of food
    (b) a rate of growth which is related to the number of individuals present, so that the growth curve becomes increasingly steep
    (c) the development of agriculture

57. (a) (i) $38 \div 18 = 2 \cdot 1$
        (ii) 60 000 000
    (b) by including more families in the survey
    (c) demography

58. (a) It caused an increase in nitrate concentrations.
    (b) (i) It would be as similar as possible to the other area, except the trees would not have been removed.
        (ii) It is important to have a control area to ensure that the factor being studied is the only cause of any changes which are recorded./It shows that the change in nitrate concentration has been caused by the removal of the trees.
    (c) Trees can recycle up to 50% of the water landing on a forest. If there are no trees, the water runs off the land into the rivers rather than evaporating from the leaves into the atmosphere.
    (d) (i) It declined.
        (ii) The nitrogen is being washed out of the soil and is not being replaced because the trees have been removed.
    (e) by the activity of denitrifying bacteria

59. (a) smallpox   (b) AIDS
    (c) whooping cough and measles
    (d) phenylketonuria   (e) malaria

60. (a) (i) CO_2 (ii) CO_2
        (iii) because there is much more CO_2 in the atmosphere than CH_4, so the CO_2 has a disproportionate effect
    (b) leakage from natural gas systems/paddy fields/decaying organic matter/intestines of many herbivores
    (c) increased sea levels/climate change/ extremes of weather/changes to ecosystems/spread of tropical diseases

Answers to Specimen Paper

SECTION A

1. C	11. B	21. A
2. B	12. C	22. B
3. A	13. B	23. A
4. C	14. D	24. B
5. D	15. C	25. D
6. B	16. C	26. A
7. C	17. C	27. D
8. A	18. B	28. C
9. D	19. A	29. A
10. D	20. D	30. B

SECTION B

1. (a) A: competitive
   B: non-competitive
   (b) The competitive inhibitor attaches to the active site (temporarily) and prevents the substrate from reacting with it. The non-competitive inhibitor attaches to another part of the enzyme and alters the shape of the active site so that the substrate does not fit properly.
   (c) substrate

2. (a) (i) CAA and CAG
       (ii) 6
   (b) They indicate where the codes for a particular protein start and stop.
   (c) (i) leucine and lysine
       (ii) proline and lysine
       (iii) substitution
       (iv) It only changes one amino acid in the protein. Deletion and insertion mutations can affect all subsequent amino acids in the chain.

3. (a) (i) first (ii) second (iii) second (iv) first
   (b) This is a stage of **meiosis** since chromosomes (not chromatids) are moving to the poles.

4. (a)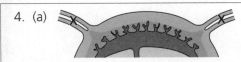
   (b) progesterone or oestrogen
   (c) because blood pressure of mother is too high/because two blood groups might be incompatible
   (d) oxygen by diffusion/glucose by active transport/antibodies by pinocytosis

5. (a)

		Blood group of recipient			
		A	B	AB	O
Blood group of donor	A	✓	✗	✓	✗
	B	✗	✓	✓	✗
	AB	✗	✗	✓	✗
	O	✓	✓	✓	✓

   (b) (i)

Gametes → ↓	A	O
A	AA	AO
O	AO	OO

       (ii) 25% (1 in 4) *(The second event is unaffected by the first event.)*
   (c) (i) Q
       (ii) S and T

6. (a) A: mitosis
       B: cell division
   (b) They lose their nucleus./They produce haemoglobin./They become biconcave in shape.
   (c) vitamin B_{12}
   (d) 100 days

7. (a) (i) It slows it down.
       (ii) medulla (oblongata)
       (iii) parasympathetic
   (b) It acts as a pacemaker.
   (c) synapses
   (d) the head
   (e)

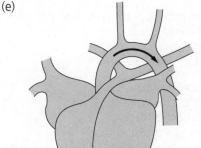

8. (a) 70 bpm *(read off right hand scale)*
   (b) (i) 4·9 litres per minute
       (ii) 8 litres per minute
   (c) heart rate from **graph 2** = 80 bpm
       ✗ should be on intersect with 30 seconds.
   (d) resting heart rate/recovery of heart rate
       after exercise

9. (a) (i) glomerulus/Bowman's capsule
       (ii) proximal (convoluted) tubule
   (b) 2·9 ± 0·1 g/l
   (c) ¾ (75%)
   (d) *Any two from:* vitamins/amino acids/water/
       salts
   (e) to reabsorb water/to increase salt
       concentration of kidney tissues

10. (a) (i)

Flask	Temperature drop (°C)
A	25
B	35
C	45

   (ii) Flask A would cool most slowly
        because it is insulated and because it
        has a relatively low surface area to
        volume ratio.
        OR
        Flask C would cool most quickly
        because it is not insulated and because
        it has a relatively high surface area to
        volume ratio.
   (b) a 100 cm³ flask which is uninsulated
   (c) hypothermia

11. (a) These numbers (4 and 3) appeared **twice**.
    (b) Numbers in the middle of the list are not
        remembered so well as those at the
        beginning and end of the list.
    (c) 7
    (d) The numbers in the STM had not been
        displaced.
    (e) chunking

12. (a) (i) cerebellum
        (ii) medulla oblongata
        (iii) limbic system
        (iv) medulla oblongata
    (b) lips, tongue, digits (fingers and thumbs).
    (c) identification

13. (a) Different concentrations of nitrate may
        have been used by the farmer.
        One crop may have been treated with
        fertiliser, the other may not.
        The nitrate concentrations in the soils
        might be different.
        Potatoes may have a higher demand for
        nitrogen than wheat.
        Potato roots may absorb more nitrogen
        than wheat roots.
    (b) Nitrate concentration might give the farmer
        an indication of the health of the crop./
        High concentrations could indicate that too
        much fertiliser had been used, and vice
        versa.
    (c) The water of the loch is fairly stationary./
        The river is constantly replenished by fresh
        (moving) water.
    (d) It can cause algal blooms/plants to thrive.
    (e) for the manufacture of protein/DNA

14. (a) the spring water and the deep well
        The spring water is coming from high
        ground which is unpolluted.
        OR
        The spring water is filtered as it flows
        through the porous rock.
        The deep well contains water which has
        been filtered through many layers of rock/
        has not been contaminated.
    (b) typhoid/cholera/dysentery
    (c) *Any two from:* The water would be cloudy/
        muddy./There would be floods of
        contaminated water./The rainfall in the
        region might decline.
    (d) People living in developed countries use
        more water for: watering gardens and
        washing cars, dishes, clothes and
        themselves.
        OR
        Factories use large volumes of water.
        OR
        Power stations use large volumes of water
        for cooling purposes.

SECTION C

One mark given for each line up to maximum indicated.
Solidus (/) indicates an alternative answer.
Brackets indicate useful information which is not essential to gain marks.

1. A (i) **Social facilitation**
   - Performance of any task usually improves in the presence of others
     e.g. athletic performance improves when spectators are present (audience effect)

   - Because there is motivation/pressure to impress others/show off

   - Because there is the additional element of competition/comparison against others
     e.g. athletic performance improves when competing with others (coactor effect)

   - Performance in more complex tasks can be inhibited by the presence of others.   (max 5)

   (ii) **Deindividuation**
   - The individual adopts behaviour of group.

   - Behaviour is atypical/not normal for that individual.

   - Behaviour is usually anti-social/bad
     e.g. football supporters will run on to pitch/fight/shout abuse in the company
     of others

   - Behaviour results from:
       attempt to impress others/showing off
       reduced risk of being caught   } *max 2*
       wishing to imitate others whom one admires.   (max 5)

1. B (i) **Carbon dioxide**
   - Gases trap heat of sun, causing global warming/rise in sea levels.

   - Volumes/concentrations increasing due to human activity
     e.g. burning/removal of forests
       Consequent reduction in photosynthesis to remove carbon dioxide
     e.g. increased use of fossil fuels
       Burning petrol in cars/burning coal and gas in power stations   (max 5)

   (ii) **Methane**
   - CH_4
   - Present in natural gas
     Released to the atmosphere from drilling/extraction processes/use of natural gas
     by humans

   - Results from metabolism of bacteria
     Bacteria are found in the soil/swamps/paddy fields.
     Bacteria are found in the stomachs/guts of many animals.   (max 5)

© Leckie & Leckie

In the following essays **2 marks** are given for the quality of the construction of the essay, in particular the **relevance** and **coherence** of the composition.

---

2. A **Structure of proteins**
- Thousands of different proteins
- Made up of 20 different amino acids
- Elements: CHON (+ S sometimes)
- Sequence of amino acids determined by sequence of bases on DNA
- Amino acids linked by peptide bonds
  to give primary structure
- Short chains called peptides
- Longer chains called polypeptides
- Linked by other bonds such as hydrogen bonds *(1 mark)* to form folds/sheets/spirals *(1 mark)* which is secondary structure *(1 mark) (any two for 2 marks)*
- Linked by other bonds (e.g. hydrogen/sulphide bonds) to form tertiary structure.   (max 8)

2. B **Functions of lipids**
- Lipids are subdivided into fats, oils, waxes and steroids *(any two)*
- Source of energy for body/respiration
- Weight for weight, two times more energy than carbohydrates (and proteins)
- Heat insulation – poor conductor of heat
- Myelin sheath – insulates nerve/causes nerve impulses to travel more quickly
- Component of cell membranes in phospholipid bilayer/cholesterol
- Needed for vitamin transport (vitamins A and D)
- Sebum in skin
  protects against infection/keeps skin supple/waterproof
- Hormones such as oestrogen/progesterone/testosterone/sex hormones are steroids
- Fat pads on feet/hands for protection   (max 8)

SOME INFORMATION AND ADVICE ON ANSWERING SECTIONS B AND C

- Spelling should be as accurate as possible, but does not have to be perfect. So *systoly* for *systole*, or *protien* for *protein*, would be marked correct. However, where two words might be confused, such as *meiosis* and *mitosis* or *glycogen* and *glucagon*, your spelling must be accurate.

- When asked for a numerical answer, make sure you state the units if units are not given, e.g. *20°C*. *20* on its own would be marked wrong.

- If a description or explanation is required, make sure you write a proper sentence or two. One-word answers will be penalised.

- It is acceptable to use well-known abbreviations such as DNA, ATP, FSH and ADH without having to write the names in full.

- Make sure your writing is legible. If an examiner can't read your answer you will get no marks.

Section C

- An answer need be no more than one page of A4.
- Clearly labelled diagrams can gain good marks.
- Keep to the subject and don't repeat yourself.
- Take time to read your answers over to make sure they make sense.

57. cont.

Graph 2 shows the distribution of family size in a random sample of 18 families in the UK.

(a) (i) Calculate the average family size (fertility rate) for the sample shown in **graph 2**.

(ii) Using this value and the data in **graph 1**, predict the population of the UK in 2100.

(b) How could this prediction be made more reliable?

(c) What term is used to describe such studies of population trends?

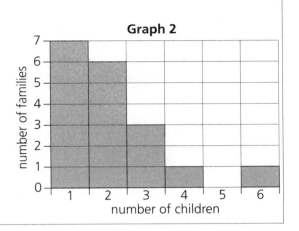

Graph 2

HUMAN INFLUENCE ON THE ENVIRONMENT

- Be aware of the impact on the environment of the increasing demand for agricultural land.
- Know the purpose and impact of the use of fertilisers and pesticides in agriculture.
- Be aware of the importance of selective breeding and genetic engineering in the production of food.
- Know that the supply of water is an increasing problem for both developed and developing countries.
- Know about the consequences of deforestation and marginal-land farming practices on the water supply cycle.
- Know about the regulatory effect of diseases, such as malaria, on populations.
- Be able to describe the use of vaccines to control many major diseases such as measles, whooping cough and smallpox.
- Know the effects of improved hygiene and sanitation on the spread of disease.
- Be able to explain how food webs are disrupted by human activities such as using chemicals to improve crop growth.
- Be aware that the loss of complexity of ecosystems leads to instability of ecosystems.
- Know how the nitrogen and carbon cycles have been disrupted by human activities.
- Know the effects of inadequate sewage treatment on the environment.
- Be aware of the consequences of global warming caused by increased concentrations of gases such as carbon dioxide and methane.

58. The graph shows the changes in nitrate concentration of river water flowing out of an area of forest where the trees have been removed.

(a) What effect did the removal of the trees have on the nitrate concentration of the river?

(b) (i) Describe the likely appearance of the area chosen as a 'control'.

(ii) Explain why a 'control' area is important.

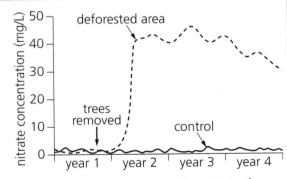

(c) The average volume of the river flowing out of the deforested area increased by 40% after the trees were removed. Suggest a reason for this change.

(d) (i) Describe the change in nitrate concentration of the river in the deforested area over the final two years of the study.

(ii) Suggest a reason for this change.

(e) Describe **one** other way in which nitrogen can be lost from a natural ecosystem.

See pages 62, 65, 68 and 69 of Leckie & Leckie's *Higher Human Biology Course Notes.*

59. For the questions which follow, take your answers from the list of human diseases and disorders given below.
(Each answer is used only once or not at all.)

**tetanus, malaria, smallpox, bubonic plague,
AIDS, phenylketonuria, measles, whooping cough**

(a) Which disease has been eradicated worldwide by an extensive vaccination programme?

(b) Name a viral disease for which there is, as yet, no vaccine.

(c) Which **two** diseases are still major childhood diseases?

(d) Which of the above is a genetic disorder brought about by a block in a metabolic pathway?

(e) Which disease is caused by a protozoan and still results in millions of deaths every year?

---

60. The figures below relate to the greenhouse gases carbon dioxide (CO_2), methane (CH_4), nitrous oxide (N_2O) and chlorofluorocarbons (CFCs).

Relative greenhouse effects per molecule of gas			
CO_2	CH_4	N_2O	CFCs
1	29	155	17 500

Contribution of gases to global warming

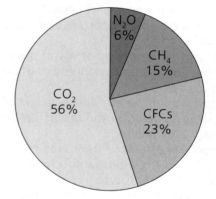

(a) (i) From the table above, which gas has the lowest greenhouse effect?
    (ii) From the pie chart, which gas contributes most to global warming?
    (iii) Give a reason for the apparent contradiction in the answers given to (i) and (ii).

(b) Give **one** example of a source of methane gas.

(c) Describe **one** possible consequence of global warming.

Specimen Paper

SECTION A
All questions in this section should be attempted.

1. What is the name of the cell structure **P** shown in the diagram below?

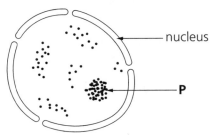

   A   ribosome
   B   lysosome
   C   nucleolus
   D   Golgi body

2. The width of a cell is 80 μm.
   1000 μm = 1 millimetre.
   What is the width of the cell expressed in metres?
   A 8×10^{-4}
   B 8×10^{-5}
   C 8×10^{-6}
   D 8×10^{-7}

3. Which of the following substances is a monosaccharide?

   A   glucose
   B   glycogen
   C   glucagon
   D   glycerol

4. The key below can be used to identify some carbohydrates.

   1 { soluble   go to 2
       insoluble   glycogen

   2 { Benedict's test positive   go to 3
       Benedict's test negative   sucrose

   3 { Barfoed's test positive   go to 4
       Barfoed's test negative   lactose

   4 { Clinistix test positive   glucose
       Clinistix test negative   fructose

According to the information provided in the key, which test can be used to distinguish between **glucose** and **lactose**?

   A   solubility
   B   Benedict's
   C   Barfoed's
   D   Clinistix

5. A stock solution contains 6% salt. Which of the following procedures would be used to produce a solution of 2% salt?

   A   Add 2 cm³ of water to 10 cm³ stock solution.
   B   Add 4 cm³ of water to 10 cm³ stock solution.
   C   Add 20 cm³ of water to 60 cm³ stock solution.
   D   Add 20 cm³ of water to 10 cm³ stock solution.

6. The graph below shows the results of an investigation into the effect of oxygen concentration on the rate of ion uptake by a cell.

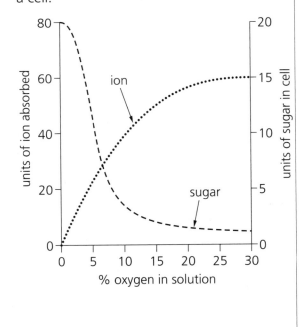

What conclusion can be drawn from the graph?

A   The sugar solution diffuses out of the cells and the ion diffuses into the cells.
B   The concentration of sugar declines as the concentration of oxygen increases.
C   The ion is absorbing oxygen from the solution and using sugar as a source of energy.
D   The cell is using sugar as a source of energy to export the ion by active transport.

7. With which type of substance is the term *active site* associated?

A   nucleic acid
B   phospholipid
C   enzyme
D   acetyl coenzyme A

8. Which of the following diagrams shows two nucleotides linked correctly?

A

B

C

D

Questions 9 and 10 refer to the diagram below which shows some of the metabolic pathways in aerobic and anaerobic respiration.

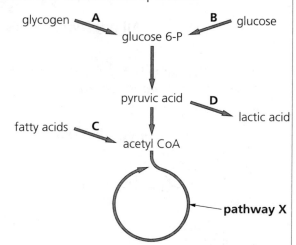

9. Which of the pathways, **A**, **B**, **C** or **D**, is purely anaerobic?

10. Which of the following correctly describes the chain of events in **pathway X**?

A   ATP is manufactured from ADP and P_i.
B   Compounds are oxidised and reduced in turn.
C   Amino acids are linked by peptide bonds in sequence.
D   Carbon atoms are removed from some compounds.

11. How many carbon atoms does one molecule of pyruvic acid contain?

A   2
B   3
C   4
D   6

12. Which of the following compounds acts as a hydrogen carrier during aerobic respiration?

A   ATP
B   ADP
C   NAD
D   ADH

13. Which line of the table correctly identifies a stage of respiration and the site at which it occurs?

	Stage	Site in mitochondrion
A	Krebs cycle	cristae
B	Krebs cycle	matrix
C	glycolysis	cristae
D	glycolysis	matrix

Questions 14 and 15 refer to a recessive genetic condition which is **not** sex-linked.

14. A man and woman, who are both heterozygous for the condition, have a normal child. What are the chances that the child is carrying the recessive allele?

   A   1 in 2
   B   1 in 3
   C   1 in 4
   D   2 in 3

15. The woman is pregnant with a second child. What are the chances that this second child will suffer from the condition?

   A   1 in 2
   B   1 in 3
   C   1 in 4
   D   2 in 3

16. What is meant by the term *artificial insemination?*

   A   the transfer of genes from bacteria to human cells
   B   the fertilisation of an egg in a glass dish
   C   the transfer of sperm to the oviduct using a syringe
   D   the injection of antibodies after infection

Questions 17 and 18 refer to the diagram of the ovary.

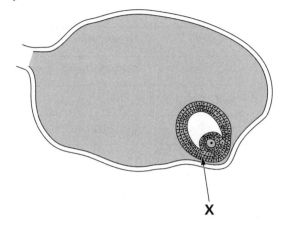

X

17. What hormone is produced by structure **X**?

   A   FSH
   B   LH
   C   oestrogen
   D   progesterone

18. After ovulation structure **X** will become

   A   a corpus callosum
   B   a corpus luteum
   C   the placenta
   D   the zygote

19. Which of the following shows the correct series of events after the formation of a zygote?

   A   cleavage → implantation → differentiation
   B   implantation → cleavage → differentiation
   C   cleavage → differentiation → implantation
   D   implantation → differentiation → cleavage

20. Which of the following is measured using a sphygmomanometer?

   A   breathing rate
   B   depth of breathing
   C   heart rate
   D   blood pressure

21. Which line of the table shows correctly an endocrine gland and a hormone which it produces?

	Gland	Hormone
A	pituitary	growth hormone
B	adrenal	ADH
C	pancreas	growth hormone
D	thyroid	TSH

22. The diagram represents a model of a human. What is the surface area to volume ratio of the model?

    A   4:1
    B   17:4
    C   9:2
    D   19:4

23. In what part of the brain are the motor, sensory and auditory centres located?

    A   cerebrum
    B   medulla oblongata
    C   cerebellum
    D   corpus callosum

Questions 24 and 25 can be answered by reference to the diagram of the neuromuscular junction shown below.

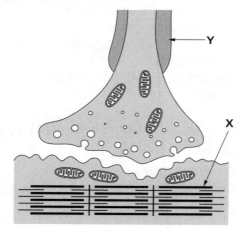

24. What is the name of the protein marked **X**?

    A   myelin
    B   myosin
    C   actin
    D   adenine

25. Which of the following is **not** a function of the area marked **Y**?

    A   to insulate the nerve fibre
    B   to protect the nerve fibre
    C   to speed up impulse transmission
    D   to store neurotransmitter

26. The principle symptom of Alzheimer's disease is

    A   memory loss
    B   paralysis
    C   high fever
    D   sickness.

27. Which of the following is an example of verbal communication?

    A   smiling
    B   sticking tongue out
    C   licking lips
    D   whispering

28. The term *carrying capacity* describes

    A   the number of items which can be stored in STM
    B   the volume of food which the stomach can hold
    C   the number of organisms a habitat can sustain
    D   the strength of muscle fibres in the arm.

29. Which of the following pairs of diseases are controlled by the vaccination of young children?

    A   measles and whooping cough
    B   measles and malaria
    C   PKU and malaria
    D   PKU and whooping cough

30. The diagram below shows the age composition of four countries.

    In which country is 18% of the population over the age of 65?

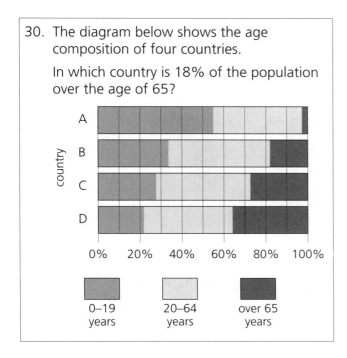

SOME INFORMATION AND ADVICE ON ANSWERING SECTION A

- The first question is often made easier to give you a good start.

- Usually, examiners ensure that there is a roughly equal balance of A, B, C and D answers.
  So make sure that you do not have a wide variation in the numbers of A, B, C and D answers.

- If you find you have answered questions with the same letter four or more times in a row, e.g. AAAA, suspect an error on your part.

- About one third of the questions will be on problem solving, e.g. questions 2, 4, 5, 6, 14, 15, 22 and 30 in this specimen paper. In these questions you are given most or all the information needed to answer the question.

- Some people advise answering the question yourself before you look at the options. This is because the options are designed to confuse!

- Never leave an answer blank. There is always a 25% chance you will get the answer right!

- You can score out wrong answers and change answers, so long as you make clear your final decision.

SECTION B
All questions in this section should be attempted.

1. The diagrams show an enzyme being inhibited by a competitive and a non-competitive inhibitor.

   A 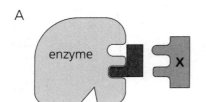   B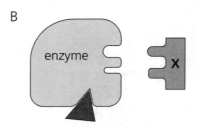

   (a) Match each type of inhibitor to the appropriate diagram. (1)

   (b) For each type of inhibitor, briefly describe how it affects the activity of the enzyme. (2)

   (c) What term is used to describe molecule **X** which would normally interact with the enzyme? (1)

2. The table below shows the mRNA codons for all the amino acids found in human proteins.

First position	Second position				Third position
	U	C	A	G	
U	phenylalanine	serine	tyrosine	cysteine	U
	phenylalanine	serine	tyrosine	cysteine	C
	leucine	serine	*Stop*	*Stop*	A
	leucine	serine	*Stop*	tryptophan	G
C	leucine	proline	histidine	arginine	U
	leucine	proline	histidine	arginine	C
	leucine	proline	glutamine	arginine	A
	leucine	proline	glutamine	arginine	G
A	isoleucine	threonine	asparagine	serine	U
	isoleucine	threonine	asparagine	serine	C
	isoleucine	threonine	lysine	arginine	A
	methionine/*Start*	threonine	lysine	arginine	G
G	valine	alanine	aspartic acid	glycine	U
	valine	alanine	aspartic acid	glycine	C
	valine	alanine	glutamic acid	glycine	A
	valine	alanine	glutamic acid	glycine	G

2. cont.
   (a) (i) From the table, the mRNA codons for histidine are CAU and CAC.
       What are the **two** mRNA codons for glutamine? (1)
       (ii) How many mRNA codons are there for leucine? (1)

   (b) A number of codons code for *start* and *stop*. What is the significance of these start/stop codons? (1)

   (c) The diagram below shows a base sequence in part of a DNA strand.

       G A C T T T

       (i) What **two** amino acids are coded for by this strand of DNA? (1)
       (ii) If adenine was replaced by guanine as a result of a mutation in the DNA strand above, which **two** amino acids would result? (1)
       (iii) What name is given to this type of mutation? (1)
       (iv) Why is this type of mutation potentially less harmful than an insertion or deletion mutation? (2)

---

3. (a) The following statements relate to the process of meiosis. For each statement state whether it applies to the **first** or **second** division of meiosis.

   (i) Homologous chromosomes pair.
   (ii) Chromatids separate.
   (iii) Chromosomes line up singly along the equator of the cell.
   (iv) Crossing-over takes place. (2)

   (b) Describe what is happening in the diagram opposite and state whether this is a stage of meiosis or mitosis. (2)

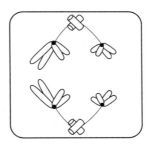

---

4. The diagram below shows a fetus developing in the uterus.

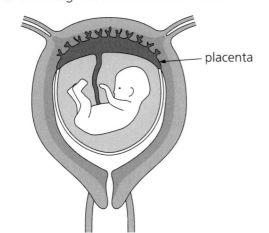

placenta

   (a) On the diagram mark with the letter **X** the place where fertilisation takes place. (1)

   (b) Name a hormone produced by the placenta. (1)

   (c) The placenta keeps the blood of the fetus separate from the blood of the mother. Give **one** reason why this is important. (1)

   (d) Name a useful substance transferred from the mother through the placenta to the fetus and state the method by which the substance crosses the membrane of the placenta. (1)

5. The table below shows which blood groups can be safely transfused from one person to another.
Some of the boxes are completed.

(a) Complete the table using ticks and crosses as shown.

		Blood group of recipient			
		A	B	AB	O
Blood group of donor	A	✓	✕	✓	✕
	B		✓		
	AB	✕		✓	
	O				✓

✓ blood compatible
✕ blood not compatible

(2)

(b) A husband and wife are both blood group **A**.
They have one child who is blood group **O**.
(i) Complete the Punnett square to show the possible genotypes of the parents' gametes and the possible genotypes of the F_1 generation.

Gametes → ↓		

(2)

(ii) The wife is pregnant with a second child. What are the chances that the second child will have the same genotype as the first child? (1)

(c) The list below contains information relating to the children and to the phenotypes of a family which might be at risk with respect to the Rhesus blood group.

P   the first child
Q   the second child

R   the mother is Rhesus positive
S   the mother is Rhesus negative
T   the father is Rhesus positive
V   the father is Rhesus negative

Select the appropriate letters to indicate:
(i) the child who would be most at risk (1)
(ii) the phenotypes of the parents which could lead to a child being at risk. (2)

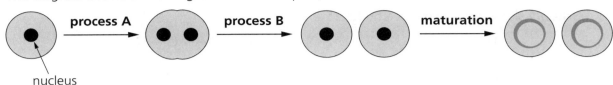

6. The diagram shows some stages in the development of red blood cells.

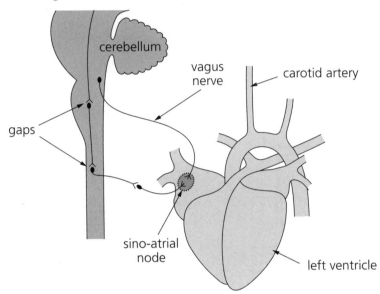

(a) Name **process A** and **process B**. (2)

(b) State a change which takes place in red blood cells as they mature. (1)

(c) Name a vitamin specifically required for the manufacture of red blood cells. (1)

(d) What is the approximate life span of a red blood cell?
Choose your answer from the list below.
1 day, 100 days, 1 month, 10 months, 1 year, 10 years (1)

7. The diagram below shows the heart and its nerve supply.

(a) (i) What effect does stimulation of the vagus nerve have on the heartbeat? (1)
(ii) To what part of the brain is the vagus nerve connected? (1)
(iii) To what branch of the nervous system does the vagus nerve belong? (1)

(b) What is the function of the sino-atrial node? (1)

(c) What is the term used to name the gaps shown between nerves? (1)

(d) To where does the carotid artery lead? (1)

(e) Draw an arrow on the aorta to show the direction of blood flow. (1)

8. The graphs below relate to the performance of a person's heart before and during exercise.

   Graph 1 shows stroke volume and heart rate.

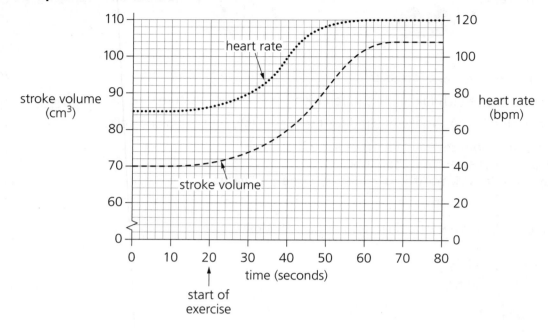

Graph 2 shows an ECG trace taken during the exercise.

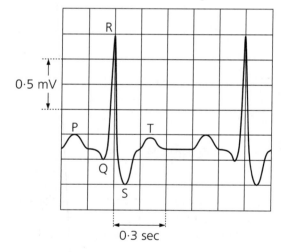

(a) From **Graph 1**, what is the heart rate of the person at rest? (1)

(b) Cardiac output is measured by multiplying the **heart rate** by the **stroke volume**.
   (i) What is the cardiac output, in litres per minute, before the start of the exercise? (1)
   (ii) What is the cardiac output, in litres per minute, 20 seconds after the start of the exercise? (1)

(c) Calculate the heart rate of the individual from **Graph 2** and mark with a cross (✕), on the appropriate line of **Graph 1**, the point at which the ECG was taken. (2)

(d) Suggest **one** feature of heart rate which would give an indication of the physical fitness of an individual. (1)

9. The graph shows the relationship between plasma concentration of glucose and the rates at which it is filtered, reabsorbed and excreted by a nephron in a healthy human.

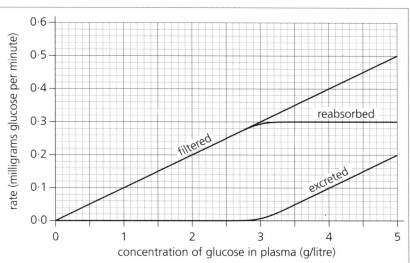

(a) (i) In what part of the nephron does filtration take place? (1)
    (ii) In what part of the nephron does most reabsorption take place? (1)

(b) At what concentration of plasma glucose does glucose first appear in the urine? (1)

(c) At a plasma concentration of 4·0 g/litre what proportion of the filtered glucose is reabsorbed? (1)

(d) Name **two** substances, other than glucose, which are reabsorbed in the kidney tubule. (1)

(e) What is the function of the Loop of Henle? (1)

---

10. An experiment was carried out to measure heat loss from three flasks containing water at 95°C. The flasks were set up as shown and left for an hour. The temperatures of the water in the flasks after an hour were 50°C, 60°C and 70°C.

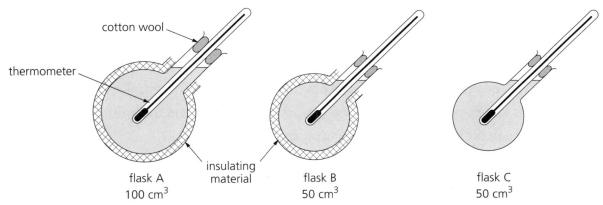

(a) (i) Match each drop in temperature with the appropriate flask by completing the table. (1)

Flask	Temperature drop (°C)
A	
B	
C	

10. cont.
    (a) (ii) Give **two** reasons for your choice in the table in part (a) (i).   (2)

    (b) Suggest a suitable control which would provide more reliable data.   (1)

    (c) What term is used to describe a drop in body temperature below 35°C?   (1)

11. An experiment was carried out to measure displacement from short-term memory (STM). A sequence of 12 random numbers was read out to a student. The student was then given a number (the probe) and asked which number in the list followed the probe. The probe only appeared once in the list.

    Example: 246873951403
    Probe 2 requires answer 4.
    Probe 7 requires answer 3.

    The experiment was repeated with 50 students, using the same sequence of numbers. The results of the experiment are shown in the graph opposite.

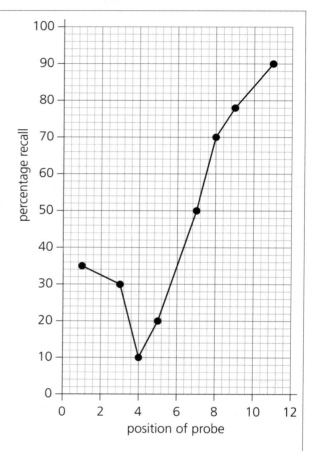

    (a) Why was percentage recall for probe positions 2, 6 and 10 not recorded on the graph?   (1)

    (b) What is the relationship between the position of the probe and the probability of recall? (1)

    (c) What is the position of the probe which would give 50% recall?   (1)

    (d) Give a reason for the improved recall when the probe is near the end of the list.   (1)

    (e) In the example given above, the first four numbers might be easily remembered as a single piece of information. What term is used to describe this process which increases the capacity of STM?   (1)

12. (a) Name the regions of the brain associated with each of the following processes. Take your answers from the list below. Each answer may be used **once**, **more than once** or **not at all**.

limbic system, cerebellum, pituitary gland, medulla oblongata, corpus callosum

    (i) fine co-ordination of muscular movement
    (ii) control of breathing rate
    (iii) memory storage
    (iv) reflex actions   (4)

  (b) With reference to the sensory homunculus shown below, which **three** parts of the body are most sensitive to touch?   (1)

  (c) A company employs a pop singer to promote its product. It is hoped the general public will imitate the pop star and buy the product.
    What term is used to describe this form of behaviour?   (1)

13. A survey of nitrate concentrations was carried out on a farm on a day in July. Nitrate concentrations in leaves of crop plants, nearby loch water and nearby river water are given in the table below.

Sample	Nitrate concentration in mg/dm³
wheat leaf	400
potato leaf	750
loch water	200
river water	100

  (a) Suggest a reason for the difference in nitrate concentration of the wheat leaf and the potato leaf.   (1)

  (b) Why might the nitrate concentration of leaves be of interest to the farmer?   (1)

  (c) Suggest a reason for the higher concentration of nitrate in the loch compared with the river.   (1)

  (d) What effect can high nitrate concentration have on lochs and rivers?   (1)

  (e) Why is it important that humans receive an adequate supply of nitrogen in their diet?   (1)

14. The diagram below shows **three** sources of drinking water near a village in a developing country.

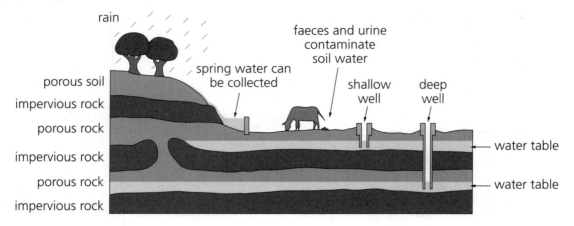

(a) Which **two** sources of water are most safe to drink?
Give a reason for each of your answers. (2)

(b) Name a human disease which is carried in water polluted by sewage. (1)

(c) If the trees were cut down, describe **two** possible effects this might have on the water supply. (2)

(d) State **two** reasons why the volume of water consumed per head of population in a developed country is much greater than in a developing country. (2)

SECTION C
**Answer BOTH questions 1 and 2, each of which has a choice.
You may use labelled diagrams where appropriate.**

1. Answer **either** A **or** B.
   A. Give an account of group behaviour under the following headings:
      (i) social facilitation (5)
      (ii) deindividuation. (5)
      (10)

   OR

   B. Give an account of the global increase in greenhouse gases under the following headings:
      (i) carbon dioxide (5)
      (ii) methane. (5)
      (10)

2. Answer **either** A **or** B.
   A. Give an account of the structure of proteins. (10)

   OR

   B. Give an account of the functions of lipids in the body. (10)

Total marks for paper **130**

Index to Words in *Higher Human Biology Course Notes*

The page numbers in this index refer to pages in the companion text to this book, Leckie & Leckie's *Higher Human Biology Course Notes* (ISBN 1-84372-144-9). Where a page number is printed in **bold**, then this page contains a definition of, or the most important information about, the word(s).